Author's N

This book features 100 influential and inspiring quotes by Albert Einstein. Undoubtedly, this collection will give you a huge boost of inspiration.

COPYRIGHT © 2021 DAVID SMITH

1

"Any fool can know. The point is to understand."

2

"The only thing that interferes with my learning is my education."

3

"A calm and modest life brings more happiness than the pursuit of success combined with constant restlessness."

4

"Intellectuals solve problems, geniuses prevent them."

5

"If you want your children to be intelligent, read them fairy tales. If you want them to be more intelligent, read them more fairy tales."

6

"He who can no longer pause to wonder and stand rapt in awe, is as good as dead; his eyes are closed."

7

"The search for truth is more precious than its possession."

8

"I always get by best with my naivety, which is 20 percent deliberate."

9

"A happy man is too satisfied with the present to dwell too much on the future."

10

"Nothing truly valuable arises from ambition or from a mere sense of duty; it stems rather from love and devotion toward men and toward objective things."

11

"Weak people revenge. Strong people forgive. Intelligent people ignore."

12

"Don't listen to the person who has the answers; listen to the person who has the questions."

13

"The only thing more dangerous than ignorance is arrogance."

14

"Long live impudence! It's my guardian angel in this world."

15

"Most people say that it is the intellect which makes a great scientist. They are wrong: it is character."

16

"The ideals which have lighted me on my way and time after time given me new courage to face life cheerfully, have been truth, goodness, and beauty."

17

"If you are out to describe the truth, leave elegance to the tailor."

18

"The value of achievement lies in the achieving."

19

"Three rules of work: Out of clutter find simplicity; From discord find harmony; In the middle of difficulty lies opportunity."

20

"On the mysterious: It is the fundamental emotion which stands at the cradle of true art and true science. He who knows it not and can no longer wonder, no longer feel amazement, is as good as dead, a snuffed-out candle."

21

"I am by heritage a Jew, by citizenship a Swiss, and by makeup a human being, and only a human being, without any special attachment to any state or national entity whatsoever."

22

"If people are good only because they fear punishment, and hope for reward, then we are a sorry lot indeed."

23

"To raise new questions, new possibilities, to regard old problems from a new angle, requires creative imagination and marks real advance in science."

24

"I do not at all believe in human freedom in the philosophical sense. Everybody acts not only under external compulsion but also in accordance with inner necessity."

25

"Any intelligent fool can make things bigger, more complex, and more violent. It takes a touch of genius, and a lot of courage, to move in the opposite direction."

26

"Unthinking respect for authority is the greatest enemy of truth."

27

"The only thing I did was this: in long intervals I have expressed an opinion on public issues whenever they appeared to me so bad and unfortunate that silence would have made me feel guilty of complicity."

28

"The value of a man should be seen in what he gives and not in what he is able to receive."

29

"When I am judging a theory, I ask myself whether, if I were God, I would have arranged the world in such a way."

30

"I believe in intuition and inspiration. At times I feel certain I am right while not knowing the reason."

31

"The only real valuable thing is intuition."

32

"The only source of knowledge is experience."

33

"If you can't explain it simply, you don't understand it well enough."

34

"It's not that I'm so smart, it's just that I stay with problems longer."

35

"Creativity is intelligence having fun."

36

"I lived in that solitude which is painful in youth, but delicious in maturity."

37

"I want to know God's thoughts; the rest are details."

38

"Make everything as simple as possible, but not simpler."

39

"The world is not dangerous because of those who do harm but because of those who look at it without doing anything."

40

"The most incomprehensible thing about the world is that it is at all comprehensible."

41

"The most powerful force in the universe is compound interest."

42

"You have to learn the rules of the game. And then you have to play better than anyone else."

43

"There are only two ways to live your life. One is as though nothing is a miracle. The other is as though everything is a miracle."

44

"No problem can be solved from the same level of consciousness that created it."

45

"Try not to become a man of success but rather to become a man of value."

46

"Once we accept our limits, we go beyond them."

47

"A little knowledge is dangerous. So is a lot."

48

"God is subtle but he is not malicious."

49

"Never give up on what you really want to do. The person with big dreams is more powerful than one with all the facts."

50

"The destiny of civilized humanity depends more than ever on the moral forces it is capable of generating."

51

"Measured objectively, what a man can wrest from Truth by passionate striving is utterly infinitesimal. But the striving frees us from the bonds of the self and makes us comrades of those who are the best and the greatest."

52

"A new type of thinking is essential if mankind is to survive and move toward higher levels."

53

"As for the search for truth, I know from my own painful searching, with its many blind alleys, how hard it is to take a reliable step, be it ever so small, towards the understanding of that which is truly important."

54

"A theory is the more impressive the greater the simplicity of its premises, the more different kinds of things it relates, and the more extended its area of applicability."

55

"Only those who attempt the absurd can achieve the impossible."

56

"The really valuable thing in the pageant of human life seems to me not the State but the creative, sentient individual, the personality; it alone creates the noble and the sublime, while the herd as such remains dull in thought and full in feeling."

57

"The pursuit of truth and beauty is a sphere of activity in which we are permitted to remain children all our lives."

58

"The man who regards his own life and that of his fellow creatures as meaningless is not merely unfortunate but almost disqualified for life."

59

"The really good music, whether of the East or of the West, cannot be analyzed."

60

"If I were not a physicist, I would probably be a musician. I often think in music. I live my daydreams in music. I see my life in terms of music."

61

"Bureaucracy is the death of all sound work."

62

"I'm doing just fine, considering that I have triumphantly survived Nazism and two wives."

63

"I have reached an age when, if someone tells me to wear socks, I don't have to."

64

"Women always worry about things that men forget; men always worry about things women remember."

65

"Any man who can drive safely while kissing a pretty girl is simply not giving the kiss the attention it deserves."

66

"There is a race between mankind and the universe. Mankind is trying to build bigger, better, faster, and more foolproof machines. The universe is trying to build bigger, better, and faster fools. So far the universe is winning."

67

"As for the words of warm praise addressed to me, I shall carefully refrain from disputing them. For who still believes that there is such a thing as genuine modesty? I should run the risk of being taken for just an old hypocrite."

68

"Science is a wonderful thing if one does not have to earn one's living at it."

69

"If we knew what it was we were doing, it would not be called research, would it?"

70

"Do not worry about your difficulties in mathematics. I can assure you mine are still greater."

71

"Two things are infinite: the universe and human stupidity; and I'm not sure about the the universe."

72

"The wireless telegraph is not difficult to understand. The ordinary telegraph is like a very long cat. You pull the tail in New York, and it meows in Los Angeles. The wireless is the same, only without the cat."

73

"Put your hand on a hot stove for a minute, and it seems like an hour. Sit with a pretty girl for an hour, and it seems like a minute. That's relativity."

74

"Too many of us look upon Americans as dollar chasers. This is a cruel libel, even if it is reiterated thoughtlessly by the Americans themselves."

75

"The devil has put a penalty on all things we enjoy in life. Either we suffer in health or we suffer in soul or we get fat."

76

"To punish me for my contempt for authority, fate made me an authority myself."

77

"A perfection of means, and confusion of aims, seems to be our main problem."

78

"Mankind invented the atomic bomb, but no mouse would ever construct a mousetrap."

79

"Stay away from negative people. They have a problem for every solution."

80

"Wisdom is not a product of schooling but of the lifelong attempt to acquire it."

81

"Man usually avoids attributing cleverness to somebody else, unless it is an enemy."

82

"Whoever is careless with the truth in small matters cannot be trusted with important matters."

83

"All of one's contemporaries and aging friends are living in a delicate balance, and one feels that one's own consciousness is no longer as brightly lit as it once was. But then, twilight with its more subdued colors has its charms as well."

84

"I soon learned to scent out what was able to lead to fundamentals and to turn aside from everything else, from the multitude of things that clutter up the mind."

85

"It would be possible to describe everything scientifically, but it would make no sense; it would be without meaning, as if you described a Beethoven symphony as a variation of wave pressure."

86

"No amount of experimentation can ever prove me right; a single experiment can prove me wrong."

87

"A new idea comes suddenly and in a rather intuitive way, but intuition is nothing but the outcome of earlier intellectual experience."

88

"As far as the laws of mathematics refer to reality, they are not certain, as far as they are certain, they do not refer to reality."

89

"The eternal mystery of the universe is its comprehensibility."

90

"The intuitive mind is a sacred gift and the rational mind is a faithful servant. We have created a society that honors the servant and has forgotten the gift."

91

"Common sense is the collection of prejudices acquired by age 18."

92

"What is right is not always popular and what is popular is not always right."

93

"I have remained a simple fellow who asks nothing of the world; only my youth is gone – the enchanting youth that forever walks on air."

94

"Compassionate people are geniuses in the art of living, more necessary to the dignity, security, and joy of humanity than the discoverers of knowledge."

95

"A hundred times every day I remind myself that my inner and outer life depend on the labours of other men, living and dead, and that I must exert myself in order to give in the same measure as I received and am still receiving."

96

"The fear of death is the most unjustified of all fears, for there's no risk of accident for someone who's dead."

97

"I have firmly resolved to bite the dust, when my time comes, with a minimum of medical assistance, and up to then I will sin to my wicked heart's content."

98

"The old who have died live on in the young ones. Don't you feel this now in your bereavement, when you look at your children?"

99

"Our death is not an end if we have lived on in our children and the younger generations. For they are us; our bodies are only wilted leaves on the tree of life."

100

"I am strongly drawn to a frugal life and am often oppressively aware that I am engrossing an undue amount of the labor of my fellow-men."

Printed in Great Britain
by Amazon